Welding Log

ISBN-13: 978-1543124361
ISBN-10: 1543124364

Date:_____

Company/Contractor/Employer/Local:

_____

Address:

_____

City, State, Zip Code:

_____

Process:

Position:

Joint Design:

Electrode/Filler:

Welder Stencil/ID: Initial:

Employed from: _____ to _____

Check One: CWI_____ Inspector_____ Supervisor_____

Signature:_____ Date:_____

Notes:

Date:_____

Company/Contractor/Employer/Local:

_____
Address:

_____
City, State, Zip Code:

_____

Process:

Position:

Joint Design:

Electrode/Filler:

Welder Stencil/ID: Initial:

Employed from: _____ to _____

Check One: CWI_____ Inspector_____ Supervisor_____

Signature:_____ Date:_____

Notes:

Date:_____

Company/Contractor/Employer/Local:

_____

Address:

_____

City, State, Zip Code:

_____

Process:

Position:

Joint Design:

Electrode/Filler:

Welder Stencil/ID: Initial:

Employed from: _____ to _____

Check One: CWI_____ Inspector_____ Supervisor_____

Signature:_____ Date:_____

Notes:

Date:_____

Company/Contractor/Employer/Local:

_____

Address:

_____

City, State, Zip Code:

_____

Process:

Position:

Joint Design:

Electrode/Filler:

Welder Stencil/ID: Initial:

Employed from: _____ to _____

Check One: CWI_____ Inspector_____ Supervisor_____

Signature:_____ Date:_____

Notes:

Date:_____

Company/Contractor/Employer/Local:

_____

Address:

_____

City, State, Zip Code:

_____

Process:

Position:

Joint Design:

Electrode/Filler:

Welder Stencil/ID: Initial:

Employed from: _____ to _____

Check One: CWI_____ Inspector_____ Supervisor_____

Signature:_____ Date:_____

Notes:

Date:_____

Company/Contractor/Employer/Local:

_____

Address:

_____

City, State, Zip Code:

_____

Process:

Position:

Joint Design:

Electrode/Filler:

Welder Stencil/ID: Initial:

Employed from: _____ to _____

Check One: CWI_____ Inspector_____ Supervisor_____

Signature:_____ Date:_____

Notes:

Date:_____

Company/Contractor/Employer/Local:

_____

Address:

_____

City, State, Zip Code:

_____

Process:

Position:

Joint Design:

Electrode/Filler:

Welder Stencil/ID: Initial:

Employed from: _____ to _____

Check One: CWI_____ Inspector_____ Supervisor_____

Signature:_____ Date:_____

Notes:

Date:_____

Company/Contractor/Employer/Local:

_____

Address:

_____

City, State, Zip Code:

_____

Process:

Position:

Joint Design:

Electrode/Filler:

Welder Stencil/ID: Initial:

Employed from: _____ to _____

Check One: CWI_____ Inspector_____ Supervisor_____

Signature:_____ Date:_____

Notes:

Date:_____

Company/Contractor/Employer/Local:

_____

Address:

_____

City, State, Zip Code:

_____

Process:

Position:

Joint Design:

Electrode/Filler:

Welder Stencil/ID: Initial:

Employed from: _____ to _____

Check One: CWI_____ Inspector_____ Supervisor_____

Signature:_____ Date:_____

Notes:

Date:_____

Company/Contractor/Employer/Local:

_____

Address:

_____

City, State, Zip Code:

_____

Process:

Position:

Joint Design:

Electrode/Filler:

Welder Stencil/ID: Initial:

Employed from: _____ to _____

Check One: CWI_____ Inspector_____ Supervisor_____

Signature:_____ Date:_____

Notes:

Date:_____

Company/Contractor/Employer/Local:

_____

Address:

_____

City, State, Zip Code:

_____

Process:

Position:

Joint Design:

Electrode/Filler:

Welder Stencil/ID: Initial:

Employed from: _____ to _____

Check One: CWI_____ Inspector_____ Supervisor_____

Signature:_____ Date:_____

Notes:

Date:_____

Company/Contractor/Employer/Local:

_____
Address:

_____
City, State, Zip Code:

_____

Process:

Position:

Joint Design:

Electrode/Filler:

Welder Stencil/ID: Initial:

Employed from: _____ to _____

Check One: CWI_____ Inspector_____ Supervisor_____

Signature:_____ Date:_____

Notes:

Date:_____

Company/Contractor/Employer/Local:

_____

Address:

_____

City, State, Zip Code:

_____

Process:

Position:

Joint Design:

Electrode/Filler:

Welder Stencil/ID: Initial:

Employed from: _____ to _____

Check One: CWI_____ Inspector_____ Supervisor_____

Signature:_____ Date:_____

Notes:

Date:_____

Company/Contractor/Employer/Local:

_____

Address:

_____

City, State, Zip Code:

_____

Process:

Position:

Joint Design:

Electrode/Filler:

Welder Stencil/ID: Initial:

Employed from: _____ to _____

Check One: CWI_____ Inspector_____ Supervisor_____

Signature:_____ Date:_____

Notes:

Date:_____

Company/Contractor/Employer/Local:

_____

Address:

_____

City, State, Zip Code:

_____

Process:

Position:

Joint Design:

Electrode/Filler:

Welder Stencil/ID: Initial:

Employed from: _____ to _____

Check One: CWI_____ Inspector_____ Supervisor_____

Signature:_____ Date:_____

Notes:

Date:_____

Company/Contractor/Employer/Local:

_____

Address:

_____

City, State, Zip Code:

_____

Process:

Position:

Joint Design:

Electrode/Filler:

Welder Stencil/ID: Initial:

Employed from: _____ to _____

Check One: CWI_____ Inspector_____ Supervisor_____

Signature:_____ Date:_____

Notes:

Date:_____

Company/Contractor/Employer/Local:

_____

Address:

_____

City, State, Zip Code:

_____

Process:

Position:

Joint Design:

Electrode/Filler:

Welder Stencil/ID: Initial:

Employed from: _____ to _____

Check One: CWI_____ Inspector_____ Supervisor_____

Signature:_____ Date:_____

Notes:

Date:_____

Company/Contractor/Employer/Local:

_____

Address:

_____

City, State, Zip Code:

_____

Process:

Position:

Joint Design:

Electrode/Filler:

Welder Stencil/ID: Initial:

Employed from: _____ to _____

Check One: CWI_____ Inspector_____ Supervisor_____

Signature:_____ Date:_____

Notes:

Date:_____

Company/Contractor/Employer/Local:

_____

Address:

_____

City, State, Zip Code:

_____

Process:

Position:

Joint Design:

Electrode/Filler:

Welder Stencil/ID: Initial:

Employed from: _____ to _____

Check One: CWI_____ Inspector_____ Supervisor_____

Signature:_____ Date:_____

Notes:

Date:_____

Company/Contractor/Employer/Local:

_____

Address:

_____

City, State, Zip Code:

_____

Process:

Position:

Joint Design:

Electrode/Filler:

Welder Stencil/ID: Initial:

Employed from: _____ to _____

Check One: CWI_____ Inspector_____ Supervisor_____

Signature:_____ Date:_____

Notes:

Date:_____

Company/Contractor/Employer/Local:

_____

Address:

_____

City, State, Zip Code:

_____

Process:

Position:

Joint Design:

Electrode/Filler:

Welder Stencil/ID: Initial:

Employed from: _____ to _____

Check One: CWI_____ Inspector_____ Supervisor_____

Signature:_____ Date:_____

Notes:

Date:_____

Company/Contractor/Employer/Local:

_____

Address:

_____

City, State, Zip Code:

_____

Process:

Position:

Joint Design:

Electrode/Filler:

Welder Stencil/ID: Initial:

Employed from: _____ to _____

Check One: CWI_____ Inspector_____ Supervisor_____

Signature:_____ Date:_____

Notes:

Date:_____

Company/Contractor/Employer/Local:

_____

Address:

_____

City, State, Zip Code:

_____

Process:

Position:

Joint Design:

Electrode/Filler:

Welder Stencil/ID: Initial:

Employed from: _____ to _____

Check One: CWI_____ Inspector_____ Supervisor_____

Signature:_____ Date:_____

Notes:

Date:_____

Company/Contractor/Employer/Local:

_____

Address:

_____

City, State, Zip Code:

_____

Process:

Position:

Joint Design:

Electrode/Filler:

Welder Stencil/ID: Initial:

Employed from: _____ to _____

Check One: CWI_____ Inspector_____ Supervisor_____

Signature:_____ Date:_____

Notes:

Date:_____

Company/Contractor/Employer/Local:

_____

Address:

_____

City, State, Zip Code:

_____

Process:

Position:

Joint Design:

Electrode/Filler:

Welder Stencil/ID: Initial:

Employed from: _____ to _____

Check One: CWI_____ Inspector_____ Supervisor_____

Signature:_____ Date:_____

Notes:

Date:_____

Company/Contractor/Employer/Local:

_____

Address:

_____

City, State, Zip Code:

_____

Process:

Position:

Joint Design:

Electrode/Filler:

Welder Stencil/ID: Initial:

Employed from: _____ to _____

Check One: CWI_____ Inspector_____ Supervisor_____

Signature:_____ Date:_____

Notes:

Date:_____

Company/Contractor/Employer/Local:

_____

Address:

_____

City, State, Zip Code:

_____

Process:

Position:

Joint Design:

Electrode/Filler:

Welder Stencil/ID: Initial:

Employed from: _____ to _____

Check One: CWI_____ Inspector_____ Supervisor_____

Signature:_____ Date:_____

Notes:

Date:_____

Company/Contractor/Employer/Local:

_____

Address:

_____

City, State, Zip Code:

_____

Process:

Position:

Joint Design:

Electrode/Filler:

Welder Stencil/ID: Initial:

Employed from: _____ to _____

Check One: CWI_____ Inspector_____ Supervisor_____

Signature:_____ Date:_____

Notes:

Date:_____

Company/Contractor/Employer/Local:

_____

Address:

_____

City, State, Zip Code:

_____

Process:

Position:

Joint Design:

Electrode/Filler:

Welder Stencil/ID: Initial:

Employed from: _____ to _____

Check One: CWI_____ Inspector_____ Supervisor_____

Signature:_____ Date:_____

Notes:

Date:_____

Company/Contractor/Employer/Local:

_____

Address:

_____

City, State, Zip Code:

_____

Process:

Position:

Joint Design:

Electrode/Filler:

Welder Stencil/ID: Initial:

Employed from: _____ to _____

Check One: CWI_____ Inspector_____ Supervisor_____

Signature:_____ Date:_____

Notes:

Date:_____

Company/Contractor/Employer/Local:

_____

Address:

_____

City, State, Zip Code:

_____

Process:

Position:

Joint Design:

Electrode/Filler:

Welder Stencil/ID: Initial:

Employed from: _____ to _____

Check One: CWI_____ Inspector_____ Supervisor_____

Signature:_____ Date:_____

Notes:

Date:_____

Company/Contractor/Employer/Local:

_____

Address:

_____

City, State, Zip Code:

_____

Process:

Position:

Joint Design:

Electrode/Filler:

Welder Stencil/ID: Initial:

Employed from: _____ to _____

Check One: CWI_____ Inspector_____ Supervisor_____

Signature:_____ Date:_____

Notes:

Date:_____

Company/Contractor/Employer/Local:

_____

Address:

_____

City, State, Zip Code:

_____

Process:

Position:

Joint Design:

Electrode/Filler:

Welder Stencil/ID: Initial:

Employed from: _____ to _____

Check One: CWI_____ Inspector_____ Supervisor_____

Signature:_____ Date:_____

Notes:

Date:_____

Company/Contractor/Employer/Local:

_____

Address:

_____

City, State, Zip Code:

_____

Process:

Position:

Joint Design:

Electrode/Filler:

Welder Stencil/ID: Initial:

Employed from: _____ to _____

Check One: CWI_____ Inspector_____ Supervisor_____

Signature:_____ Date:_____

Notes:

Date:_____

Company/Contractor/Employer/Local:

_____

Address:

_____

City, State, Zip Code:

_____

Process:

Position:

Joint Design:

Electrode/Filler:

Welder Stencil/ID: Initial:

Employed from: _____ to _____

Check One: CWI_____ Inspector_____ Supervisor_____

Signature:_____ Date:_____

Notes:

Date:_____

Company/Contractor/Employer/Local:

_____

Address:

_____

City, State, Zip Code:

_____

Process:

Position:

Joint Design:

Electrode/Filler:

Welder Stencil/ID: Initial:

Employed from: _____ to _____

Check One: CWI_____ Inspector_____ Supervisor_____

Signature:_____ Date:_____

Notes:

Date:_____

Company/Contractor/Employer/Local:

_____

Address:

_____

City, State, Zip Code:

_____

Process:

Position:

Joint Design:

Electrode/Filler:

Welder Stencil/ID: Initial:

Employed from: _____ to _____

Check One: CWI_____ Inspector_____ Supervisor_____

Signature:_____ Date:_____

Notes:

Date:_____

Company/Contractor/Employer/Local:

_____
Address:

_____
City, State, Zip Code:

_____

Process:

Position:

Joint Design:

Electrode/Filler:

Welder Stencil/ID: Initial:

Employed from: _____ to _____

Check One: CWI_____ Inspector_____ Supervisor_____

Signature:_____ Date:_____

Notes:

Date:_____

Company/Contractor/Employer/Local:

_____

Address:

_____

City, State, Zip Code:

_____

Process:

Position:

Joint Design:

Electrode/Filler:

Welder Stencil/ID: Initial:

Employed from: _____ to _____

Check One: CWI_____ Inspector_____ Supervisor_____

Signature:_____ Date:_____

Notes:

Date:_____

Company/Contractor/Employer/Local:

_____

Address:

_____

City, State, Zip Code:

_____

Process:

Position:

Joint Design:

Electrode/Filler:

Welder Stencil/ID: Initial:

Employed from: _____ to _____

Check One: CWI_____ Inspector_____ Supervisor_____

Signature:_____ Date:_____

Notes:

Date:_____

Company/Contractor/Employer/Local:

_____

Address:

_____

City, State, Zip Code:

_____

Process:

Position:

Joint Design:

Electrode/Filler:

Welder Stencil/ID: Initial:

Employed from: _____ to _____

Check One: CWI_____ Inspector_____ Supervisor_____

Signature:_____ Date:_____

Notes:

Date:_____

Company/Contractor/Employer/Local:

_____

Address:

_____

City, State, Zip Code:

_____

Process:

Position:

Joint Design:

Electrode/Filler:

Welder Stencil/ID: Initial:

Employed from: _____ to _____

Check One: CWI_____ Inspector_____ Supervisor_____

Signature:_____ Date:_____

Notes:

Date:_____

Company/Contractor/Employer/Local:

_____

Address:

_____

City, State, Zip Code:

_____

Process:

Position:

Joint Design:

Electrode/Filler:

Welder Stencil/ID: Initial:

Employed from: _____ to _____

Check One: CWI_____ Inspector_____ Supervisor_____

Signature:_____ Date:_____

Notes:

Date:_____

Company/Contractor/Employer/Local:

_____

Address:

_____

City, State, Zip Code:

_____

Process:

Position:

Joint Design:

Electrode/Filler:

Welder Stencil/ID: Initial:

Employed from: _____ to _____

Check One: CWI_____ Inspector_____ Supervisor_____

Signature:_____ Date:_____

Notes:

Date:_____

Company/Contractor/Employer/Local:

_____

Address:

_____

City, State, Zip Code:

_____

Process:

Position:

Joint Design:

Electrode/Filler:

Welder Stencil/ID: Initial:

Employed from: _____ to _____

Check One: CWI_____ Inspector_____ Supervisor_____

Signature:_____ Date:_____

Notes:

Date:_____

Company/Contractor/Employer/Local:

_____

Address:

_____

City, State, Zip Code:

_____

Process:

Position:

Joint Design:

Electrode/Filler:

Welder Stencil/ID: Initial:

Employed from: _____ to _____

Check One: CWI_____ Inspector_____ Supervisor_____

Signature:_____ Date:_____

Notes:

Date:_____

Company/Contractor/Employer/Local:

_____

Address:

_____

City, State, Zip Code:

_____

Process:

Position:

Joint Design:

Electrode/Filler:

Welder Stencil/ID: Initial:

Employed from: _____ to _____

Check One: CWI_____ Inspector_____ Supervisor_____

Signature:_____ Date:_____

Notes:

Date:_____

Company/Contractor/Employer/Local:

_____

Address:

_____

City, State, Zip Code:

_____

Process:

Position:

Joint Design:

Electrode/Filler:

Welder Stencil/ID: Initial:

Employed from: _____ to _____

Check One: CWI_____ Inspector_____ Supervisor_____

Signature:_____ Date:_____

Notes:

Date:_____

Company/Contractor/Employer/Local:

_____

Address:

_____

City, State, Zip Code:

_____

Process:

Position:

Joint Design:

Electrode/Filler:

Welder Stencil/ID: Initial:

Employed from: _____ to _____

Check One: CWI_____ Inspector_____ Supervisor_____

Signature:_____ Date:_____

Notes:

Date:_____

Company/Contractor/Employer/Local:

_____

Address:

_____

City, State, Zip Code:

_____

Process:

Position:

Joint Design:

Electrode/Filler:

Welder Stencil/ID: Initial:

Employed from: _____ to _____

Check One: CWI_____ Inspector_____ Supervisor_____

Signature:_____ Date:_____

Notes:

Date:_____

Company/Contractor/Employer/Local:

_____

Address:

_____

City, State, Zip Code:

_____

Process:

Position:

Joint Design:

Electrode/Filler:

Welder Stencil/ID: Initial:

Employed from: _____ to _____

Check One: CWI_____ Inspector_____ Supervisor_____

Signature:_____ Date:_____

Notes:

Date:_____

Company/Contractor/Employer/Local:

_____

Address:

_____

City, State, Zip Code:

_____

Process:

Position:

Joint Design:

Electrode/Filler:

Welder Stencil/ID: Initial:

Employed from: _____ to _____

Check One: CWI_____ Inspector_____ Supervisor_____

Signature:_____ Date:_____

Notes:

Date:_____

Company/Contractor/Employer/Local:

_____

Address:

_____

City, State, Zip Code:

_____

Process:

Position:

Joint Design:

Electrode/Filler:

Welder Stencil/ID: Initial:

Employed from: _____ to _____

Check One: CWI_____ Inspector_____ Supervisor_____

Signature:_____ Date:_____

Notes:

Date:_____

Company/Contractor/Employer/Local:

_____

Address:

_____

City, State, Zip Code:

_____

Process:

Position:

Joint Design:

Electrode/Filler:

Welder Stencil/ID: Initial:

Employed from: _____ to _____

Check One: CWI_____ Inspector_____ Supervisor_____

Signature:_____ Date:_____

Notes:

Date:_____

Company/Contractor/Employer/Local:

_____

Address:

_____

City, State, Zip Code:

_____

Process:

Position:

Joint Design:

Electrode/Filler:

Welder Stencil/ID: Initial:

Employed from: _____ to _____

Check One: CWI_____ Inspector_____ Supervisor_____

Signature:_____ Date:_____

Notes:

Date:_____

Company/Contractor/Employer/Local:

_____

Address:

_____

City, State, Zip Code:

_____

Process:

Position:

Joint Design:

Electrode/Filler:

Welder Stencil/ID: Initial:

Employed from: _____ to _____

Check One: CWI_____ Inspector_____ Supervisor_____

Signature:_____ Date:_____

Notes:

Date:_____

Company/Contractor/Employer/Local:

_____

Address:

_____

City, State, Zip Code:

_____

Process:

Position:

Joint Design:

Electrode/Filler:

Welder Stencil/ID: Initial:

Employed from: _____ to _____

Check One: CWI_____ Inspector_____ Supervisor_____

Signature:_____ Date:_____

Notes:

Date:_____

Company/Contractor/Employer/Local:

_____

Address:

_____

City, State, Zip Code:

_____

Process:

Position:

Joint Design:

Electrode/Filler:

Welder Stencil/ID: Initial:

Employed from: _____ to _____

Check One: CWI_____ Inspector_____ Supervisor_____

Signature:_____ Date:_____

Notes:

Date:_____

Company/Contractor/Employer/Local:

_____

Address:

_____

City, State, Zip Code:

_____

Process:

Position:

Joint Design:

Electrode/Filler:

Welder Stencil/ID: Initial:

Employed from: _____ to _____

Check One: CWI_____ Inspector_____ Supervisor_____

Signature:_____ Date:_____

Notes:

Date:_____

Company/Contractor/Employer/Local:

_____
Address:

_____
City, State, Zip Code:

_____

Process:

Position:

Joint Design:

Electrode/Filler:

Welder Stencil/ID: Initial:

Employed from: _____ to _____

Check One: CWI_____ Inspector_____ Supervisor_____

Signature:_____ Date:_____

Notes:

Date:_____

Company/Contractor/Employer/Local:

_____

Address:

_____

City, State, Zip Code:

_____

Process:

Position:

Joint Design:

Electrode/Filler:

Welder Stencil/ID: Initial:

Employed from: _____ to _____

Check One: CWI_____ Inspector_____ Supervisor_____

Signature:_____ Date:_____

Notes:

Date:_____

Company/Contractor/Employer/Local:

_____

Address:

_____

City, State, Zip Code:

_____

Process:

Position:

Joint Design:

Electrode/Filler:

Welder Stencil/ID: Initial:

Employed from: _____ to _____

Check One: CWI_____ Inspector_____ Supervisor_____

Signature:_____ Date:_____

Notes:

Date:_____

Company/Contractor/Employer/Local:

_____

Address:

_____

City, State, Zip Code:

_____

Process:

Position:

Joint Design:

Electrode/Filler:

Welder Stencil/ID: Initial:

Employed from: _____ to _____

Check One: CWI_____ Inspector_____ Supervisor_____

Signature:_____ Date:_____

Notes:

Date:_____

Company/Contractor/Employer/Local:

_____

Address:

_____

City, State, Zip Code:

_____

Process:

Position:

Joint Design:

Electrode/Filler:

Welder Stencil/ID: Initial:

Employed from: _____ to _____

Check One: CWI_____ Inspector_____ Supervisor_____

Signature:_____ Date:_____

Notes:

Date:_____

Company/Contractor/Employer/Local:

_____

Address:

_____

City, State, Zip Code:

_____

Process:

Position:

Joint Design:

Electrode/Filler:

Welder Stencil/ID: Initial:

Employed from: _____ to _____

Check One: CWI_____ Inspector_____ Supervisor_____

Signature:_____ Date:_____

Notes:

Date:_____

Company/Contractor/Employer/Local:

_____

Address:

_____

City, State, Zip Code:

_____

Process:

Position:

Joint Design:

Electrode/Filler:

Welder Stencil/ID: Initial:

Employed from: _____ to _____

Check One: CWI_____ Inspector_____ Supervisor_____

Signature:_____ Date:_____

Notes:

Date:_____

Company/Contractor/Employer/Local:

_____

Address:

_____

City, State, Zip Code:

_____

Process:

Position:

Joint Design:

Electrode/Filler:

Welder Stencil/ID: Initial:

Employed from: _____ to _____

Check One: CWI_____ Inspector_____ Supervisor_____

Signature:_____ Date:_____

Notes:

Date:_____

Company/Contractor/Employer/Local:

_____

Address:

_____

City, State, Zip Code:

_____

Process:

Position:

Joint Design:

Electrode/Filler:

Welder Stencil/ID: Initial:

Employed from: _____ to _____

Check One: CWI_____ Inspector_____ Supervisor_____

Signature:_____ Date:_____

Notes:

Date:_____

Company/Contractor/Employer/Local:

_____

Address:

_____

City, State, Zip Code:

_____

Process:

Position:

Joint Design:

Electrode/Filler:

Welder Stencil/ID: Initial:

Employed from: _____ to _____

Check One: CWI_____ Inspector_____ Supervisor_____

Signature:_____ Date:_____

Notes:

Date:_____

Company/Contractor/Employer/Local:

_____

Address:

_____

City, State, Zip Code:

_____

Process:

Position:

Joint Design:

Electrode/Filler:

Welder Stencil/ID: Initial:

Employed from: _____ to _____

Check One: CWI_____ Inspector_____ Supervisor_____

Signature:_____ Date:_____

Notes:

Date:_____

Company/Contractor/Employer/Local:

_____

Address:

_____

City, State, Zip Code:

_____

Process:

Position:

Joint Design:

Electrode/Filler:

Welder Stencil/ID: Initial:

Employed from: _____ to _____

Check One: CWI_____ Inspector_____ Supervisor_____

Signature:_____ Date:_____

Notes:

Date:_____

Company/Contractor/Employer/Local:

_____

Address:

_____

City, State, Zip Code:

_____

Process:

Position:

Joint Design:

Electrode/Filler:

Welder Stencil/ID: Initial:

Employed from: _____ to _____

Check One: CWI_____ Inspector_____ Supervisor_____

Signature:_____ Date:_____

Notes:

Date:_____

Company/Contractor/Employer/Local:

_____

Address:

_____

City, State, Zip Code:

_____

Process:

Position:

Joint Design:

Electrode/Filler:

Welder Stencil/ID: Initial:

Employed from: _____ to _____

Check One: CWI_____ Inspector_____ Supervisor_____

Signature:_____ Date:_____

Notes:

Date:_____

Company/Contractor/Employer/Local:

_____
Address:

_____
City, State, Zip Code:

_____

Process:

Position:

Joint Design:

Electrode/Filler:

Welder Stencil/ID: Initial:

Employed from: _____ to _____

Check One: CWI_____ Inspector_____ Supervisor_____

Signature:_____ Date:_____

Notes:

Date:_____

Company/Contractor/Employer/Local:

_____

Address:

_____

City, State, Zip Code:

_____

Process:

Position:

Joint Design:

Electrode/Filler:

Welder Stencil/ID: Initial:

Employed from: _____ to _____

Check One: CWI_____ Inspector_____ Supervisor_____

Signature:_____ Date:_____

Notes:

Date:_____

Company/Contractor/Employer/Local:

_____

Address:

_____

City, State, Zip Code:

_____

Process:

Position:

Joint Design:

Electrode/Filler:

Welder Stencil/ID: Initial:

Employed from: _____ to _____

Check One: CWI_____ Inspector_____ Supervisor_____

Signature:_____ Date:_____

Notes:

Date:_____

Company/Contractor/Employer/Local:

_____

Address:

_____

City, State, Zip Code:

_____

Process:

Position:

Joint Design:

Electrode/Filler:

Welder Stencil/ID: Initial:

Employed from: _____ to _____

Check One: CWI_____ Inspector_____ Supervisor_____

Signature:_____ Date:_____

Notes:

Date:_____

Company/Contractor/Employer/Local:

_____

Address:

_____

City, State, Zip Code:

_____

Process:

Position:

Joint Design:

Electrode/Filler:

Welder Stencil/ID: Initial:

Employed from: _____ to _____

Check One: CWI_____ Inspector_____ Supervisor_____

Signature:_____ Date:_____

Notes:

Date:_____

Company/Contractor/Employer/Local:

_____

Address:

_____

City, State, Zip Code:

_____

Process:

Position:

Joint Design:

Electrode/Filler:

Welder Stencil/ID: Initial:

Employed from: _____ to _____

Check One: CWI_____ Inspector_____ Supervisor_____

Signature:_____ Date:_____

Notes:

Date:_____

Company/Contractor/Employer/Local:

_____

Address:

_____

City, State, Zip Code:

_____

Process:

Position:

Joint Design:

Electrode/Filler:

Welder Stencil/ID: Initial:

Employed from: _____ to _____

Check One: CWI_____ Inspector_____ Supervisor_____

Signature:_____ Date:_____

Notes:

Date:_____

Company/Contractor/Employer/Local:

_____

Address:

_____

City, State, Zip Code:

_____

Process:

Position:

Joint Design:

Electrode/Filler:

Welder Stencil/ID: Initial:

Employed from: _____ to _____

Check One: CWI_____ Inspector_____ Supervisor_____

Signature:_____ Date:_____

Notes:

Date:_____

Company/Contractor/Employer/Local:

_____

Address:

_____

City, State, Zip Code:

_____

Process:

Position:

Joint Design:

Electrode/Filler:

Welder Stencil/ID: Initial:

Employed from: _____ to _____

Check One: CWI_____ Inspector_____ Supervisor_____

Signature:_____ Date:_____

Notes:

Date:_____

Company/Contractor/Employer/Local:

_____

Address:

_____

City, State, Zip Code:

_____

Process:

Position:

Joint Design:

Electrode/Filler:

Welder Stencil/ID: Initial:

Employed from: _____ to _____

Check One: CWI_____ Inspector_____ Supervisor_____

Signature:_____ Date:_____

Notes:

Date:_____

Company/Contractor/Employer/Local:

_____
Address:

_____
City, State, Zip Code:

_____

Process:

Position:

Joint Design:

Electrode/Filler:

Welder Stencil/ID: Initial:

Employed from: _____ to _____

Check One: CWI_____ Inspector_____ Supervisor_____

Signature:_____ Date:_____

Notes:

Date:_____

Company/Contractor/Employer/Local:

_____

Address:

_____

City, State, Zip Code:

_____

Process:

Position:

Joint Design:

Electrode/Filler:

Welder Stencil/ID: Initial:

Employed from: _____ to _____

Check One: CWI_____ Inspector_____ Supervisor_____

Signature:_____ Date:_____

Notes:

Date:_____

Company/Contractor/Employer/Local:

_____

Address:

_____

City, State, Zip Code:

_____

Process:

Position:

Joint Design:

Electrode/Filler:

Welder Stencil/ID: Initial:

Employed from: _____ to _____

Check One: CWI_____ Inspector_____ Supervisor_____

Signature:_____ Date:_____

Notes:

Date:_____

Company/Contractor/Employer/Local:

_____

Address:

_____

City, State, Zip Code:

_____

Process:

Position:

Joint Design:

Electrode/Filler:

Welder Stencil/ID: Initial:

Employed from: _____ to _____

Check One: CWI_____ Inspector_____ Supervisor_____

Signature:_____ Date:_____

Notes:

Date:_____

Company/Contractor/Employer/Local:

_____

Address:

_____

City, State, Zip Code:

_____

Process:

Position:

Joint Design:

Electrode/Filler:

Welder Stencil/ID: Initial:

Employed from: _____ to _____

Check One: CWI_____ Inspector_____ Supervisor_____

Signature:_____ Date:_____

Notes:

Date:_____

Company/Contractor/Employer/Local:

_____

Address:

_____

City, State, Zip Code:

_____

Process:

Position:

Joint Design:

Electrode/Filler:

Welder Stencil/ID: Initial:

Employed from: _____ to _____

Check One: CWI_____ Inspector_____ Supervisor_____

Signature:_____ Date:_____

Notes:

Date:_____

Company/Contractor/Employer/Local:

_____

Address:

_____

City, State, Zip Code:

_____

Process:

Position:

Joint Design:

Electrode/Filler:

Welder Stencil/ID: Initial:

Employed from: _____ to _____

Check One: CWI_____ Inspector_____ Supervisor_____

Signature:_____ Date:_____

Notes:

Date:_____

Company/Contractor/Employer/Local:

_____

Address:

_____

City, State, Zip Code:

_____

Process:

Position:

Joint Design:

Electrode/Filler:

Welder Stencil/ID: Initial:

Employed from: _____ to _____

Check One: CWI_____ Inspector_____ Supervisor_____

Signature:_____ Date:_____

Notes:

Date:_____

Company/Contractor/Employer/Local:
_____

Address:
_____

City, State, Zip Code:

_____

Process:

Position:

Joint Design:

Electrode/Filler:

Welder Stencil/ID: Initial:

Employed from: _____ to _____

Check One: CWI_____ Inspector_____ Supervisor_____

Signature:_____ Date:_____

Notes:

Date:_____

Company/Contractor/Employer/Local:

_____

Address:

_____

City, State, Zip Code:

_____

Process:

Position:

Joint Design:

Electrode/Filler:

Welder Stencil/ID: Initial:

Employed from: _____ to _____

Check One: CWI_____ Inspector_____ Supervisor_____

Signature:_____ Date:_____

Notes:

Date:_____

Company/Contractor/Employer/Local:

_____

Address:

_____

City, State, Zip Code:

_____

Process:

Position:

Joint Design:

Electrode/Filler:

Welder Stencil/ID: Initial:

Employed from: _____ to _____

Check One: CWI_____ Inspector_____ Supervisor_____

Signature:_____ Date:_____

Notes:

Date:_____

Company/Contractor/Employer/Local:

_____

Address:

_____

City, State, Zip Code:

_____

Process:

Position:

Joint Design:

Electrode/Filler:

Welder Stencil/ID: Initial:

Employed from: _____ to _____

Check One: CWI_____ Inspector_____ Supervisor_____

Signature:_____ Date:_____

Notes:

Date:_____

Company/Contractor/Employer/Local:
_____

Address:
_____

City, State, Zip Code:

_____

Process:

Position:

Joint Design:

Electrode/Filler:

Welder Stencil/ID: Initial:

Employed from: _____ to _____

Check One: CWI_____ Inspector_____ Supervisor_____

Signature:_____ Date:_____

Notes:

Date:_____

Company/Contractor/Employer/Local:

_____

Address:

_____

City, State, Zip Code:

_____

Process:

Position:

Joint Design:

Electrode/Filler:

Welder Stencil/ID: Initial:

Employed from: _____ to _____

Check One: CWI_____ Inspector_____ Supervisor_____

Signature:_____ Date:_____

Notes:

Date:_____

Company/Contractor/Employer/Local:

_____

Address:

_____

City, State, Zip Code:

_____

Process:

Position:

Joint Design:

Electrode/Filler:

Welder Stencil/ID: Initial:

Employed from: _____ to _____

Check One: CWI_____ Inspector_____ Supervisor_____

Signature:_____ Date:_____

Notes:

Date:_____

Company/Contractor/Employer/Local:

_____

Address:

_____

City, State, Zip Code:

_____

Process:

Position:

Joint Design:

Electrode/Filler:

Welder Stencil/ID: Initial:

Employed from: _____ to _____

Check One: CWI_____ Inspector_____ Supervisor_____

Signature:_____ Date:_____

Notes:

www.ingramcontent.com/pod-product-compliance
Lightning Source LLC
Chambersburg PA
CBHW051813170526
45167CB00005B/1996